INTRODUCTION

Our Nation is one of a vast array of actors in a complex, volatile, and unpredictable security environment. Globalization and the proliferation of technology mean we face threats across a wide spectrum and competition across all domains. We're confronted by ever-evolving adversaries ranging from one person with a single interconnected computer to sophisticated capable militaries and everything in between. We're also challenged by the shear pace of change among our adversaries fueled by profound information and technology diffusion worldwide. As stated by the Chief of Staff of the Air Force in the *Global Vigilance, Global Reach and Global Power For Our Nation* vision, "despite the best analyses and projections by national security experts, the time and place of the next crisis are never certain and are rarely what we expect." Success and the guaranty of security in this dynamic environment require that we both take lessons learned from the last decade of conflict and creatively visualize the future strategic landscape. It's in this space, between learning from the past and keeping an open eye to the future, where we find opportunity.

The focused and balanced investments of the Air Force Fiscal Year 2015 S&T Program are hedges against the unpredictable future and provide pathways to a flexible, precise and lethal force at a relatively low cost in in relation to the return on investment. The Undersecretary of Defense for Acquisition, Technology and Logistics recently reminded us that complacency now and in the future is simply not an option. Maintaining, and even expanding, our technological advantage is vital to ensuring sustained freedom of access and action in air, space and cyberspace.

AIR FORCE FISCAL YEAR 2015 S&T PROGRAM

The Air Force as a whole had to make difficult trades between force structure (capacity), readiness, and modernization (capability) in the Service's Fiscal Year 2015 President's Budget submission to recover from budget uncertainty over the two previous fiscal years. The Air Force Fiscal Year 2015 President's Budget request for S&T is approximately $2.1 billion, which includes nearly $178 million in support of devolved programs consisting of High Energy Laser efforts and the University Research Initiative. This year's Air Force S&T budget request represents a decrease of $141 million or a 6.2% decrease from the Fiscal Year 2014 President's Budget request, a slightly larger reduction as compared to the overall Air Force topline reduction. This budget request re-balances basic research spending as part of the overall portfolio to increase emphasis on conducting technology demonstrations. The Air Force was able to reduce funding in the aerospace systems and materials areas while still advancing capabilities for the Air Force and the Department of Defense (DoD) by smartly leveraging research being conducted by the Defense Advanced Research Projects Agency (DARPA) in the hypersonics area.

We've learned a great deal over the last decade. The dedicated scientists and engineers of the Air Force Research Laboratory (AFRL) have successfully supported warfighters during conflicts in Iraq, Afghanistan, and North Africa through the rapid development of systems and capabilities including persistent intelligence, surveillance, and reconnaissance (ISR); data fusion and integration from multiple sensors; and near real-time monitoring of some orbiting U.S. and commercial spacecraft assets. With the pivot to the Pacific as outlined in the *Defense Strategic Guidance*, we must continue to evolve and advance "game-changing" and enabling technologies which can transform the landscape of how the Air Force flies, fights and wins against the high-end threats in contested environments envisioned in the future.

In close coordination with the requirements, intelligence and acquisition communities, we have structured our Air Force Fiscal Year 2015 S&T Program to address the highest priority needs of the Air Force across the near-, mid- and far-term; execute a balanced and integrated program that is responsive to Air Force core missions; and advance critical technical competencies needed to address the full range of product and support capabilities. The Air Force continues to focus efforts to deliberately align S&T planning, technology transition planning, development planning and early systems engineering. The linkages between these planning activities are critical to initiating acquisition programs with more mature technologies and credible cost estimates, and we are institutionalizing these linkages in Air Force policy. Air Force S&T provides critical inputs at several phases of the Chief of Staff of the Air Force's *Air Force 2023* strategic planning effort including helping to shape the "realm of the possible" when envisioning long term strategy, offering technologies to expand the strategic viewpoint and identifying potential solutions to requirements and capability gaps. Our forthcoming updated Air Force S&T strategy focuses on investing in S&T for the future, as well as leverages our organic capacity, and the capacity of our partners (domestic and international), to integrate existing capabilities and mature technologies into innovative, affordable, and sustainable solutions. This flexible strategy provides us the technological agility to adapt our S&T Program to dynamic strategic, budgetary, and technology environments and will shape prioritized actionable S&T plans.

NEAR TERM TECHNOLOGY TRANSITION

The Air Force continues to move our Flagship Capability Concept (FCC) projects toward transition to the warfighter. A well-defined scope and specific objectives desired by a Major Command (MAJCOM) are key factors in commissioning this type of an Air Force-level technology demonstration effort. The technologies are matured by the Air Force Research Laboratory with the intent to transition to the acquisition community for eventual deployment to an end user. These

FCCs are sponsored by the using MAJCOM and are vetted through the S&T Governance Structure and Air Force Requirements Oversight Council to ensure they align with Air Force strategic priorities. In Fiscal Year 2014, the Air Force successfully completed and transitioned the Selective Cyber Operations Technology Integration (SCOTI) FCC and will continue work on the High Velocity Penetrating Weapon (HVPW) and Precision Airdrop (PAD) FCCs.

AFRL delivered the SCOTI FCC to the Air Force Life Cycle Management Center (AFLCMC) in September 2013, on time, on budget and within specification. SCOTI consists of cyber technologies capable of affecting multiple nodes for the purposes of achieving a military objective and gaining cyberspace superiority. SCOTI's robust, modular architecture provides vital extensibility to allow cyber warriors to keep pace with rapidly evolving threats. AFLCMC is evaluating the delivered SCOTI architecture for integration with operational cyber mission software to directly meet the needs of a major capability area in the Air Force Cyberspace Superiority Core Function Master Plan. By successfully meeting the requirements of the stakeholder-approved Technology Transition Plan, SCOTI is the first FCC to transition and will serve as a baseline for current and future integrated cyber tools to provide needed effects for the warfighter.

The HVPW FCC was established to demonstrate critical technologies to reduce the technical risk for a new generation of penetrating weapons to defeat difficult, hard targets. This FCC matures technologies that can be applied to the hard target munitions acquisition including guidance and control, terminal seeker, fuze, energetic materials and warhead case design. This effort develops improved penetration capability of hard, deep targets containing high strength concrete with up to 2,500 feet per second (boosted velocity) impact in a GPS-degraded environment. This technology will demonstrate penetration capability of a 5,000 pound-class gravity weapon with a 2,000 pound weapon thus enabling increased loadout for bombers and fighters. Tests will demonstrate complete warhead functionality, and are scheduled to be completed the end of September 2014.

The PAD FCC was commissioned in response to a request from the Commander of Air Mobility Command for technologies to improve airdrop accuracy and effectiveness while minimizing risk to our aircrews. To date, PAD FCC efforts have focused on: early systems engineering analysis to determine major error sources, data collection, flying with crews, wind profiling, bundle tracking, and designing modeling and simulation activities. The Air Force Research Laboratory completed the bundle tracker development in Fiscal Year 2013 and in Fiscal Year 2014 began wind profile sensor development.

GAME-CHANGING TECHNOLOGIES

The Air Force S&T Program provides technology options to enable operations in anti-access, area-denial environments and transform the way we fly, fight and win in air, space and cyberspace. To illustrate how, I will highlight some of our efforts in game-changing and enabling technology areas:

Hypersonics

Speed provides options for engagement of time sensitive targets in anti-access/area-denial environments, and improves the survivability of Air Force systems. Hypersonic speed weapons are also a force multiplier as fewer are required to defeat difficult targets and fewer platforms are required from greater standoff distances. The Air Force S&T community continues to execute the high speed technology roadmaps developed with industry over the last three years. We are also building on the success of the X-51A Waverider scramjet engine hypersonic demonstrator, which on 1 May 2013 reached an approximate Mach Number of 5.1 during its fourth and final flight. The Air Force has focused multi-faceted, phased investments in game-changing technology for survivable, time-critical strike in the near term and a penetrating regional intelligence, surveillance, and reconnaissance (ISR) and strike aircraft in the far term.

The near term strike effort is the High Speed Strike Weapon (HSSW) program. This effort will mature cruise missile technology to address many of those items necessary to realize a missile in the hypersonic speed regime including: modeling and simulation; ramjet/scramjet propulsion; high temperature materials; guidance, navigation, and control; seekers and their required apertures; warhead and subsystems; thermal protection and management; manufacturing technology; and compact energetic booster technologies.

The Air Force conducts research and development in all aspects of hypersonic technologies in partnership with NASA, DARPA, and industry/academic sectors. The HSSW program will include two parallel integrated technology demonstration efforts to leverage DARPA's recent experience in hypersonic technologies that are relevant to reduce risk in key areas. One of the demonstrations will be a tactically-relevant demonstration of an air breathing missile technology that is compatible with Air Force 5th generation platforms including geometric and weight limits for internal B-2 Spirit bomber carriage and external F-35 Lightening II fighter carriage. This demonstration will build on the X-51 success and will include a tactically compliant engine start capability and launch from a relevant altitude.

For the other demonstration, the Air Force and DARPA will seek to develop technologies and demonstrate capabilities that will enable transformational changes in prompt, survivable, long-range strike against using the Tactical Boost Glide (TBG) concept. The objective of the TBG effort is to develop and demonstrate the critical technologies that will enable an air launched tactical range, hypersonic boost-glide missile. Both efforts will build upon experience gained through recent hypersonic vehicle development and demonstration efforts supported by DARPA and the Air Force. These demonstrations are traceable to an operationally relevant weapon that could be launched from existing aircraft. Technology and concepts from these efforts will provide options for an operational weapon system for rapidly and effectively prosecuting targets in highly contested environments.

Autonomy

Analysis of these future operating environments has also led the Air Force to invest in game-changing advances in autonomous systems. Autonomous systems can extend human reach by providing potentially unlimited persistent capabilities without degradation due to fatigue or lack of attention. The Air Force S&T Program is developing technologies that realize true autonomous capabilities including those that advance the state-of-the-art in machine intelligence, decision-making, and integration with the warfighter to form effective human-machine teams.

The greater use of autonomous systems increases the capability of U.S. forces to execute well within the adversaries' decision loops. Human decision-makers intelligently integrated into autonomous systems enable the right balance of human and machine capability to meet Air Force challenges in the future. The Air Force S&T Program invests in the development of technologies to enable warfighters and machines to work together, with each understanding mission context, sharing understanding and situation awareness, and adapting to the needs and capabilities of the other. The keys to maximizing this human-machine interaction are: instilling confidence and trust among the team members; understanding of each member's tasks, intentions, capabilities and progress; and ensuring effective and timely communication. All of which must be provided within a flexible architecture for autonomy, facilitating different levels of authority, control and collaboration. Current research is focused on understanding human cognition and applying these concepts to machine learning. Efforts develop efficient interfaces for an operator to supervise multiple unmanned air systems (UAS) platforms and providing the ISR analyst with tools to assist identifying, tracking, targets of interest.

Autonomy also allows machines to synchronize activity and information. Systems that coordinate location, status, mission intent, and intelligence and surveillance data can provide redundancy, increased coverage, decreased costs and/or increased capability. Research efforts are developing control software to enable multiple, small UASs to coordinate mission tasking with other air systems or with ground sensors. Other research efforts are developing munition sensors and guidance systems that will increase operator trust, validation, and flexibility while capitalizing on the growing ability of munitions to autonomously search a region of interest, provide additional situational awareness, plan optimum flight paths, de-conflict trajectories, optimize weapon-to-target orientation, and cooperate to achieve optimum effects.

Finally, before any system is fielded, adequate testing must be conducted to demonstrate that it meets requirements and will operate as intended. As technologies with greater levels of autonomy mature, the number of test parameters will increase exponentially. Due to this increase, it will be impractical to verify and validate autonomous system performance, cost-effectively, using current methods. The Air Force is developing test techniques that verify the decision-making and logic of the system and validate the system's ability to appropriately handle unexpected situations. Efforts are focused at the software-level and build to overall system to verify codes are valid and trustworthy. The Air Force will demonstrate the tools needed to ensure autonomous systems operate safely and effectively in unanticipated and dynamic environments.

Directed Energy

With a uniquely focused directorate within AFRL, the Air Force is in a leading position in the game-changing area of directed energy. These technologies, including high powered microwave (HPM) and high energy lasers (HELs), can provide distinctive and revolutionary capabilities to several Air Force and joint mission areas. Laser technologies are rapidly evolving for infrared seeker jamming, secure communications in congested and jammed spectrum environments, space situational awareness, and vastly improved ISR and target identification capabilities at ever increasing ranges. To get HELs to a weapon system useful to the Air Force, our S&T program invests in research in laser sources from developing narrow line width fiber lasers to scaling large numbers of fiber lasers with DARPA and MDA. Since HEL devices are not sufficient for a weapon, the Air Force directed energy research also includes beam control, atmospheric compensation, acquisition, pointing, tracking, laser effects, and physics based end-to-end modeling and simulation. The Air Force also funds the High Energy Laser Joint Technology Office (HEL JTO) which supports all of the services by being the key motivator of high power laser devices such as the successful 100 kilowatt, lab-scale Joint High Power Solid State Laser (JHPSSL) and other funding many smaller successes. The current primer program, which is jointly funded with core Army and Air Force funds, is the Robust Electric Laser Initiative (RELI). The initiative funds efforts to develop designs for efficient and weaponizable solid state lasers with options leading to a 100 kilowatt laser device.

Our HPM S&T will complement kinetic weapons to engage multiple targets, neutralizing communication systems, computers, command and control nodes, and other electronics, with low collateral damage for counter-anti-access/area denial in future combat situations. The Air Force is using the results of from the highly successful Counter-Electronics High Power Microwave Advanced Missile Project (CHAMP) Joint Capabilities Technology Demonstration (JCTD) to inform an effort known as Non-Kinetic Counter Electronics (NKCE). NKCE is currently in pre-Alternative of Alternatives (AoA) phase, with an AoA potentially starting in Fiscal Year 2015. The AoA will examine the cost and performances for kinetic, non-kinetic, and cyber options for air superiority and seeks to have a procured and operational weapon system to support the targets and requirements of the Combatant Commanders in the mid-2020 time frame. In parallel, the Air Force

S&T Program is continuing HPM research and development to provide a more capable and smaller counter-electronics system that can fit onto a variety of platforms.

The DoD directed energy research community is highly integrated and the Air Force leverages the work of other agencies. For example, the Air Force is working with the Missile Defense Agency on integrated electro-optical/infrared pulsed-laser targeting to enhance situational awareness and increase survivability by enabling the use of legacy weapons in the 2016 timeframe. In addition, the Air Force is partnering with DARPA on the Demonstrator Laser Weapon System, a ground-based fully integrated laser weapon system demonstration over the next two fiscal years and an Air-to-Air Defensive Weapon Concept.

Fuel Efficiency Technologies

For the longer term reduction in energy demand, the Air Force is investing in the development of adaptive turbine engine technologies which have the potential to reduce fuel consumption while also increasing capability in anti-access/area denial environments through increased range and time-on-station. The Air Force has several priority efforts as part of the DoD's Versatile Advanced Affordable Turbine Engine (VAATE) technology program. VAATE is a coordinated Army, Navy, and Air Force plan initiated in 2003 to develop revolutionary advances in propulsion system performance, fuel efficiency and affordability for the DoD's turbine engine powered air platforms.

The initial effort, Adaptive Versatile Engine Technology (ADVENT), began in fiscal year 2007 and is set to complete this year. General Electric is currently in final testing of the ADVENT engine technologies which include a next generation high pressure ratio core and an adaptive fan in a third stream engine architecture.

The Adaptive Engine Technology Development (AETD) program, our accelerated follow-on adaptive engine effort for the combat Air Force, is progressing very well. The objective of

AETD is to fully mature adaptive engine technologies for low risk transition to multiple combat aircraft alternatives ready for fielding as soon as the early 2020's. The effort will deliver a preliminary prototype engine design, substantiated by major hardware demonstrations, that can be tailored to specific applications when the DoD is ready to launch new development programs. The overarching goal of AETD is to mature adaptive engine technologies so that these programs can launch with significantly lower risk than previous propulsion development programs.

The High Energy Efficient Turbine Engine (HEETE) S&T effort is our flagship large engine effort under the VAATE technology program. The HEETE effort's primary objective is to demonstrate engine technologies that enable a 35% fuel efficiency improvement versus the VAATE year 2000 baseline, or at least 10% beyond current VAATE technology capabilities being demonstrated in the ADVENT program.

The Air Force Research Laboratory and industry have conducted a number of HEETE payoff studies that show significant potential benefits to future transport and ISR aircraft (e.g., 18% to 30% increase in strategic transport range, 45% to 60% increase in tactical transport radius, and 37% to 75% increase in ISR UAV loiter time). A study of Air Force's fleet fuel usage showed that introduction of HEETE-derived engines into the mobility and the tanker fleet would enable fuel savings of approximately 203 million gallons per year by the mid-2030's.

Investments in these efforts help us reduce energy demand, bridge the "valley of death" between S&T and potential acquisition programs, and help maintain the U.S. industrial technological edge and lead in turbine engines.

ENABLING TECHNOLOGIES

In addition to these game-changing technologies, the Air Force S&T Program also invests in many enabling technologies to facilitate major advances and ensure maximum effectiveness in the near-, mid-, and far term:

Cyber

Operations in cyberspace magnify military effects by increasing the efficiency and effectiveness of air and space operations and by helping to integrate capabilities across all domains. However, the cyberspace domain is increasingly contested and/or denied and the Air Force faces risks from malicious insiders, insecure supply chains, and increasingly sophisticated adversaries. Fortunately, cyberspace S&T can provide assurance, resilience, affordability, and empowerment to enable the Air Force's assured cyber advantage.

In 2012, the Air Force developed *Cyber Vision 2025* which described the Air Force vision and blueprint for cyber S&T spanning cyberspace, air, space, command and control, intelligence, and mission support. *Cyber Vision 2025* provides a long-range vision for cyberspace to identify and analyze current and forecasted capabilities, threats, vulnerabilities and consequences across core Air Force missions in order to identify key S&T gaps and opportunities. The Air Force's cyber S&T investments for Fiscal Year 2015 are aligned to the four themes identified in *Cyber Vision 2025*: Mission Assurance, Agility and Resilience, Optimized Human-Machine Systems, and Foundations of Trust.

Air Force S&T efforts in Mission Assurance seek to ensure survivability and freedom of action in contested and denied environments through enhanced cyber situational awareness for air, space, and cyber commanders. Current research efforts seek to provide dynamic, real-time mapping and analysis of critical mission functions onto cyberspace. This analysis includes the cyber situation awareness functions of monitoring the health and status of cyber assets, and extends to capture how missions flow through cyberspace. This work seeks to provide commanders with the ability to recognize attacks and prioritize defensive actions to protect assets supporting critical missions. Other research efforts develop techniques to measure and assess the effects of cyber operations and integrate them with cross-domain effects to achieve military objectives.

Research in Agility and Survivability develops rapid and unpredictable maneuver capabilities to disrupt the adversaries' cyber "kill chain" along with their planning and decision-making processes and hardening cyber elements to improve the ability to fight through, survive, and rapidly recover from attacks. Air Force S&T efforts are creating dynamic, randomizable, reconfigurable architectures capable of autonomously detecting compromises, repairing and recovering from damage, and evading threats in real-time. Cyber resiliency is enhanced through an effective mix of redundancy, diversity, and distributed functionality that leverages advances in virtualization and cloud technologies.

The Air Force works to maximize the human and machine potential through the measurement of physiological, perceptual, and cognitive states to enable personnel selection, customized training, and (user, mission, and environment) tailored augmented cognition. S&T efforts develop visualization technologies to enable a global common operational picture (COP) of complex cyber capabilities that can be readily manipulated to support Air Force mission-essential functions (MEFs). Other efforts seek to identify the critical human skills and abilities that are the foundation for superior cyber warriors and develop a realistic distributed network training environment integrated with new individualized and continuous learning technologies.

The Air Force is developing secure foundations of computing to provide operator trust in Air Force weapon systems that include a mix of embedded systems, customized and militarized commercial systems, commercial off-the-shelf (COTS) equipment, and unverified hardware and software that is developed outside the United States. Research into formal verification and validation of complex, large scale, interdependent systems as well as vulnerability analysis, automated reverse engineering, and real-time forensics tools will enable designers to quantify the level of trust in various components of the infrastructure and to understand the risk these components pose to the execution of critical mission functions. Efforts to design and build secure hardware will provide a secure root-of-trust and enable a more intelligent mixing of government off-the-shelf (GOTS) and COTS components based on the systems' security requirements.

Cognitive Electronic Warfare

With the highly contested future EW environment, we have focused S&T efforts on creating the ability to rapidly respond to threats. This is accomplished by developing the analytic ability to understand a complex threat environment and determine the best combination of techniques across all available platforms. In addition, leveraging cognitive and autonomy concepts improves the cycle time between emergence of a threat and development of an effective response. This system-of-systems solution approach is implemented in a physics based interactive simulation capability to evaluate novel concepts. The Air Force is also developing technologies to enhance survivability and improve situational awareness in the electro-optical (EO)/infrared (IR) and radio frequency (RF) warning and countermeasures area. New electronic components (antennas, amplifiers, processors) will improve the ability to detect threats with emphasis on advanced processing and software to assess threats in a crowded RF environment. This includes solutions to detect and defeat infrared and optical threats. These will enable protection against autonomous seekers using multi-spectral tracking.

Space Situational Awareness / Space Control

The ability to counter threats, intentional or unintentional, in the increasingly congested and contested space domain begins with Space Situational Awareness (SSA). The SSA S&T investments needed to maintain our core Space Superiority and Command and Control missions in such an environment are substantial and include research in Assured Recognition and Persistent Tracking of Space Objects, Characterization of Space Objects and Events, Timely and Actionable Threat Warning and Assessment, and Effective Decision Support through Data Integration and

Exploitation. The Air Force works across these areas in cooperation with the DoD, intelligence community, and industry.

To help build a holistic national SSA capability, the Air Force's S&T investment is designed to exploit our in-house expertise to innovate in areas with short-, mid- and long-term impact that are not already being addressed by others. Examples include working with Federally Funded Research and Development Centers (FFRDCs) and academia to attack the deep space uncorrelated target association problem to improve custody of space objects and reduce the burden on the space surveillance network; better conjunction assessment and re-entry estimation algorithms to reduce collision probabilities and unnecessary maneuvers; and infrared star catalog improvement to ease observation calibrations. These products have recently transitioned to national SSA capabilities. Advanced component technologies developed with industry include visible focal plane arrays, deployable baffles and lenses to meet performance, and cost and weight requirements for future space-based surveillance systems.

As part of the Air Force Research Laboratory's long history of proving new technologies in relevant environments, the Automated Navigation and Guidance Experiment for Local Space (ANGELS) program examines techniques to provide a clearer picture of the environment around our vital space assets through safe, automated spacecraft operations above Geosynchronous Earth Orbit (GEO). Equipped with significant detection, tracking and characterization technology, ANGELS will launch in 2014. It will maneuver around its booster's upper stage and explore increased levels of automation in mission planning and execution, enabling more timely and complex operations with reduced footprint. Additional indications and warning work focuses on change detection and characterization technologies to provide key observables that improve response time and efficacy.

Satellite Resilience

Our Nation and our military are heavily dependent on space capabilities. With an operational space domain that is becoming increasingly congested, competitive and contested, the Air Force has seen the need for development of technologies to increase resilience of our space capabilities. The satellites upon which we rely so heavily must be able to avoid or survive threats, both man-made and natural, and to operate through and subsequently quickly recover should threat or environmental effects manifest. To this end, the Air Force S&T Program has increased technological investment in tactical sensing and threat warning, reactive satellite control, and hardening.

Satellites today are equipped with a wide range of sensors, that, if exploited in new ways and/or coupled with new hosted threat sensing technologies could yield significant increases to tactical sensing and threat warning. The Air Force pursues a range of internally-focused health and status sensing (e.g. structural integrity, thermal, cyber) and externally focused object or phenomena sensing (e.g. space environment, threat sensing, directed energy detection) technologies, and a range of data fusion approaches to maximize the timeliness and confidence of that warning. While tactical warning is vital, it is only immediately helpful when a satellite is able to tactically respond in some way to avoid a threat or minimize its effects. Any choice of a response requires some means of reconciling warning with viable courses of action available. The Air Force focuses on efforts specifically dedicated to tailoring satellite control based on tactical warning inputs. Finally, hardening technologies refers to a range of both passive and active capabilities that, when selected and executed, could result in threat avoidance, lessening their effects or recovering lost capability more quickly. For example, for particular types of threats, dynamic configuration changes, optical protection, cyber quarantine, dynamic thermal management or possibly maneuvers might achieve the desired protection.

Precision Navigation and Timing (PNT)

Most U.S. weapon systems rely on the Global Positioning System (GPS) satellites to provide the required position navigation and timing (PNT) to function properly. This reliance has created a vulnerability which is being exploited by our adversaries through development of jammers to degrade access to the GPS signals. For success in the long term, Air Force S&T is improving the robustness of military GPS receivers and also developing several non-GPS based alternative capabilities including exploitation of other satellite navigation constellations, use of new signals of opportunity, and incorporation of additional sensors such as star trackers and terrain viewing optical systems. These receivers provide new navigation options with different accuracy depending on available sensors and computational power. Rapid progress is being made on advanced Inertial Measurement Units based on cold atom technologies. These units have the potential to provide accurate PNT for extended periods without any external update. Together, these approaches will provide future options to enable the Air Force mission to continue in contested and denied environments.

Assured Communications

Assured communications are critical to the warfighter in all aspects of the Air Force core missions. The Air Force S&T Program is developing technologies to counter threats to mission performance, such as spectrum congestion and jamming, and to maintain or increase available bandwidth through access to new portions of the radio frequency spectrum, alleviating pressure on DoD spectrum allocations. Future ability to use new spectrum will increase DoD communications architecture capacity and affordability, by requiring fewer expensive, high capacity gateways. Additional bandwidth will allow improved anti-jam communications performance and higher frequency communications, which will reduce scintillation losses for nuclear command and control

(C2). The performance enhancements would directly improve the ability of remotely-piloted aircraft to transmit images and data (ISR) and improve command and control assurance.

Efforts in Assured Communications include the Future Space Communications effort which includes research to characterize and provide new spectrum for future military space communications through the W/V-band Space Communications Experiment (WSCE). WSCE will characterize and model the atmospheric effects of upper V-band and W-band (71-76 GHz and 81-86 GHz) signal transmission. Space-based data collection and atmospheric attenuation model development is necessary to provide the statistics necessary to design a future satellite communications architecture that will allow use of the currently empty V- and W-band spectrum.

Long Range Sensing

For the past decade the Air Force has provided near persistent ISR for Combatant Commanders conducting operations in the uncontested air environments of Iraq and Afghanistan. We do not see the appetite for ISR waning in the future. However, the ability to perform effective sensing in anti-access/area denial and contested environments is threatened by many new and different challenges rarely seen during the past 10 years of permissive environment operations. In the past, airborne collection platforms conducted airborne ISR outside of the lethal range of air defense systems. Today, however, the modern and evolving foreign Integrated Air Defense Systems (IADS) of our adversaries have increased lethality and significantly improved engagement capabilities which will force ISR aircraft to fly at longer stand-off distances. The effectiveness of current precision weapons will be reduced with distance limiting the ability to accurately detect, identify and geo-locate targets.

The Air Force S&T Program is focused on significantly improving our sensing ability to adequately address the challenges of extended range ISR collection. The efforts include: 1) next generation RF sensing for contested spectrum environments in which long stand-off sensing is primarily focused on all-weather ISR using traditional active radar modes at ranges of greater than 100 miles; 2) passive RF Sensing in which signals of opportunity are exploited to detect, identify and locate targets through the use of passive multi-mode and distributed multi-static techniques; 3) laser radar sensing focused on enhancing target identification through the use of synthetic aperture laser radar and also addressing high resolution wide-area three dimensional imaging through advancements in direct detection ladar; and 4) passive EO/IR sensing to enhance capabilities to detect and track difficult targets, improve target identification at long standoff ranges and perform material identification through advancing hyperspectral and stand-off high resolution imaging technology.

Live, Virtual, and Constructive (LVC)

The Air Force continues to develop and demonstrate technologies for Live, Virtual, and Constructive (LVC) operations to maintain combat readiness. The training need for LVC is real while training costs are increasing and threat environments are complex. In particular, realistic training for anti-access/area-denial environments is not available. During a recent demonstration of LVC capability for tactical forces at Shaw AFB, South Carolina, AFRL LVC research capability was integrated in operations with an F-16 Unit Training Device (a virtual simulator) to simultaneously interoperate with a mix of live F-16 aircraft, other virtual simulations, and high fidelity computer-generated constructive players. This mix of players enabled the real time and realistic portrayal and interaction of other strike package assets and aggressor aircraft with a level of complexity that could not be achieved if limited to live assets, given the expense and availability of them to support the scenarios. LVC S&T has the capability to provide greater focused training for our warfighters across a range of operational domains such as tactical air, special operations, cyber, ISR, and C2. The Air Force is exploring a 5th generation LVC Proof of Concept set of demonstrations that would validate the requirements for a formal program of record for LVC.

Basic Research

The development of revolutionary capabilities requires the careful investment in foundational science to generate new knowledge. Our scientists discover the potential military utility of these new ideas and concepts, develop this understanding to change the art-of-the-possible and then transition the S&T for further use. Air Force basic research sits at the center of an innovation network that tracks the best S&T in the DoD, with our partners in the Army, the Navy, DARPA, and the Defense Threat Reduction Agency (DTRA), while monitoring the investments and breakthroughs of the NSF, NASA, NIST, and the Department of Energy. Air Force scientists and engineers watch and collaborate with the best universities and research centers from around the world in open, publishable research that cuts across multiple scientific disciplines aligned to military needs.

For example, Air Force basic research played a role in the Air Force's successful CHAMP technology demonstration discussed earlier. While the CHAMP demonstration required extensive applied research and advanced technology development, fundamental basic research investment in both supercomputers and computational mathematics provided a virtual prototyping capability called Improved Concurrent Electromagnetic Particle-In-Cell (ICEPIC) for directed energy concepts to Air Force researchers. This allowed new ideas to be studied effectively and affordably on the computer without costly manufacture for every iteration of the technology. Virtual prototyping was a critical enabling technology, and resulted from nearly two decades of steady, targeted investments in fundamental algorithms that then transitioned to a capability driving technology development in Air Force laboratories and in industry.

Manufacturing Technologies

A key cross-cutting enabling technology area is in developing materials, processes, and advanced manufacturing technologies for all systems including aircraft, spacecraft, missiles, rockets, ground-based systems and their structural, electronic and optical components. The fiscal year 2015 Air Force S&T Program emphasizes materials work from improved design and manufacturing processes to risk reduction through assessing manufacturing readiness.

The Air Force's investment in additive manufacturing technologies offers new and innovative approaches to the design and manufacture of Air Force and DoD systems. Additive manufacturing, or the process of joining materials to make objects from 3D model data layer by layer, changes the conventional approach to design, enabling a more direct design to requirements. As opposed to subtractive processes like machining, additive manufacturing offers a whole new design realm in which geometric complexity is not a constraint and material properties can be specifically located where needed. As with the insertion of all advanced materials and processes, the Air Force strives to ensure appropriate application and proper qualification of additive manufacturing for warfighter safety and system performance.

Currently, the Air Force is invested in more than a dozen programs ranging from assisting in major high-Technology Readiness Level (TRL) qualification programs to mid-TRL process improvement programs, to low-TRL process modeling and simulation programs. Overall, we have established a strategic program to quantify risk for implementation and to advance the understanding of processing capabilities. We have identified multiple technical areas that require Air Force investment and are developing an initiative that integrates pervasive additive manufacturing technologies across Air Force sectors, spanning multiple material classes from structural, metallic applications to functional, electronic needs.

The Air Force leverages its additive manufacturing resources and interests with the Administration's National Network for Manufacturing Innovation (NNMI) to support the acceleration of additive manufacturing technologies to the U.S. manufacturing sector to increase domestic competitiveness. In fiscal year 2013, the Air Force played a key role in supporting the NNMI National Additive Manufacturing Innovation Institute called "America Makes." The Air Force, on behalf of the Office of the Secretary of Defense, led an interagency effort , which included DoD, DOE, DOC/NIST, NASA, and NSF, to launch a $69 million public-private partnership in Additive Manufacturing.

Cooperatively working with the private partner team lead, the Air Force helped "America Makes" achieve significant accomplishments in its first year. After opening it headquarters in Youngstown, Ohio in September 2012, the "America Makes" consortium has grown to approximately 80 member organizations consisting of manufacturing companies, universities, community colleges, and non-profit organizations. A shared public-private leadership governance structure, organizational charter, and intellectual property strategy were implemented and two project calls were launched in Additive Manufacturing and 3D printing technology research, discovery, creation, and innovation. So far, more than 20 projects totaling approximately $29M and involving more than 75 partners have been started covering a broad set of priorities including advances in materials, design and manufacturing processes, equipment, qualification and certification, and knowledge base development. "America Makes" serves as an example for future NNMI institutes and the Air Force has provided support to establish two additional DoD sponsored institutes of manufacturing innovation.

The Air Force Manufacturing Technology program continues to lead the way in developing methods and tools for Manufacturing Readiness Assessments and continues to lead assessments on new technology, components, processes, and subsystems to identify manufacturing maturity and associated risk. Increasing numbers of weapon system prime contractors and suppliers have integrated Manufacturing Readiness into their culture which aids in product and process transition and implementation, resulting in reduced cost, schedule and performance risk. Benefits from the advanced manufacturing propulsion initiative continue to accrue in the form of reduced turbine engine cost and weight through advanced manufacturing of light weight castings and ceramic composites and improved airfoil processing. Advanced next generation radar and coatings affordability projects continue to reduce cost and manufacturing risk to systems such as the F-22 and F-35 aircraft. The Air Force Manufacturing Technology investment continues to make a significant impact on the F-35 program in particular, driving down life cycle costs by over $3 billion, with a number of ongoing projects that will benefit multiple F-35 program Integrated Product Teams.

The Air Force is also leveraging basic research efforts to improve sustainment of legacy systems. The "Digital Twin" concept combines the state-of-the-art in computational tools, advanced sensors, and novel algorithms to create a digital model of every platform in the fleet. Imagine a world where instead of using fleet averages for the maintenance and sustainment of an airframe, there is a computer model of each plane that records all the data from each flight, integrates the stress of the flights into the history of the actual materials on the platform, and continually checks the health of vital components. Thus, the computer model mimics all the missions of the physical asset, thereby allowing us to do maintenance exactly when required. This is the airplane equivalent of individualized medicine, making sure that each individual asset of the Air Force is set to operate at peak performance. Interdisciplinary basic research in material science, fundamental studies in new sensors and novel inquiry into new, transformational computer architecture enable the Digital Twin concepts. These foundational studies are tightly integrated with applied research, both in the Air Force Research Laboratory as well as efforts in NASA, to drive forward the S&T to permit breakthroughs in affordable sustainment.

RAPID INNOVATION PROGRAM AND SMALL BUSINESS INNOVATION RESEARCH

The Air Force recognizes small businesses are critical to our defense industrial base and essential to our Nation's economy. The U.S. relies heavily on innovation through research and development as the small businesses continue to be a major driver of high-technology innovation and economic growth in the U.S. We continue to engage small businesses through the Rapid Innovation Program, and the Small Business Innovation Research (SBIR) and Small Business Technology Transfer (STTR) programs.

The Rapid Innovation Program has been an excellent means for the Air Force to communicate critical needs and solicit vendors to respond with innovative technology solutions. The program provides a vehicle for businesses, especially small businesses, to easily submit their innovative technologies where they feel it will best meet military needs. The Air Force benefits from the ability to evaluate proposed innovative technologies against critical needs, and selecting the most compelling for contract award. The response to the program has been overwhelming, and instrumental to the transition of capability by small businesses. Over the last three years, the Air Force has received over 2,200 white paper submissions from vendors offering solutions to critical Air Force needs. We have awarded over 60 projects directly to small businesses and anticipate awarding another 25 by the end of this fiscal year.

Projects from the Fiscal Year 2011 Rapid Innovation Program are now maturing and showing great promise. For example, one project developed a handheld instrument for quality assurance of surface preparation processes used in manufacturing of the F-35 aircraft. Current F-35 aircraft manufacturing processes require manual testing of 30,000 nut plates on each plane to ensure correct bonding of materials. The current failure rate is averaging 1% or 300 nut plates. Each failure requires individual re-preparation and re-bonding with supervisory oversight. The Rapid Innovation Program project handheld device will significantly reduce the failure rate of adhesively bonded nut plates. In turn, this will reduce rework and inspection costs, increase aircraft availability, assist Lockheed Martin in achieving its target production rate, and reduce repetitive injury claims from employees. Lockheed Martin has been very closely monitoring this technology and will be completing a return-on-investment review in the coming months following prototype evaluation.

The Air Force continues to collaborate with other Federal agencies and Air Force acquisition programs to streamline our SBIR and STTR processes. We are also collaborating with the Air Force's Small Business office (SAF/SB) to implement the provisions of the reauthorization and to assist in maximizing small business opportunities in government contracts while enhancing the impact and value of small businesses.

For example, to improve the effectiveness of SBIR investments, the Air Force Research Laboratory has started to strategically bundle, coordinate, and align Air Force SBIR topics against top Air Force priorities identified by Air Force Program Executive Officers (PEO). In the Fall of 2013, the Laboratory began a pilot effort with the Air Force Program Executive Officer for Space to focus the combined investments of approximately 45 SBIR Phase I awards and 15 Phase II SBIR awards on the identified, top priority challenge of transforming our military space-based PNT capabilities.

In conjunction with this strategic initiative, the Air Force is also energizing efforts to seek out and attract non-traditional participants, which are small businesses with skills, knowledge and abilities relevant to the bundled topics, in SBIR awards but who, for various reasons, do not routinely participate in the SBIR proposal process. This strategic concentration of small business innovation against top priorities will ultimately enhance the transitioning of small business innovation, raise the visibility and importance of those investments, and take advantage of the nation's small business innovation. If proven successful, the Air Force will begin to institutionalize it as a model for organizing and aligning SBIR topics against other top priority issues.

One recent SBIR project developed innovative low profile and conformal antennas to allow air platforms, including small Remotely Piloted Aircraft (RPA), to operate more aerodynamically and ground vehicles to operate more covertly in areas where Improvised Explosive Devices (IEDs) are a threat. The wideband low profile antenna assembly for vehicle Counter Radio Controlled IED Electronic Warfare (CREW) systems operates efficiently from VHF to S-band, and at a height of less than 3 inches, greatly reduces visual signature. The wideband conformal antenna technologies developed for RPA systems operate from UHF through S-band and minimize the number of required antennas, significantly reducing weight and aerodynamic drag.

WORLD CLASS WORKFORCE

Maintaining our U.S. military's decisive technological edge requires an agile, capable workforce that leads cutting-edge research, explores emerging technology areas, and promotes innovation across government, industry and academia. Nurturing our current world class workforce and the next generations of science, technology, engineering, and mathematics (STEM) professionals is an Air Force, DoD and national concern. We must be able to recruit, retain and develop a capable STEM workforce in the face of worldwide competition for the same talent.

The Air Force continues to focus on developing technical experts and leaders who can provide the very best research and technical advice across the entire lifecycle of our systems, from acquisition, test, deployment and sustainment. After yielding success since 2011, the original *Bright Horizons, the Air Force STEM Workforce Strategic Roadmap*, is currently being updated with new goals and objectives to reflect the current environment. The Air Force has also developed a soon-to-be-released *Engineering Enterprise Strategic Plan* aimed at recruiting, developing and retaining the scientist and engineer talent to meet the future need of the Air Force.

The increased Laboratory hiring and personnel management authorities and flexibilities provided by the Congress over the last several years have done much to improve our ability to attract the Nation's best talent. The Air Force is currently developing implementation plans for the authorities most recently provided in the Fiscal Year 2014 National Defense Authorization Act. The ability to manage Laboratory personnel levels according to budget will allow us to be more agile and targeted in hiring for new and emerging research areas. The Air Force Research Laboratory recruits up-and-coming, as well as seasoned, scientists and engineers, including continuing a vibrant relationship with Historically Black Colleges and Universities and Minority Serving Institutions (HBCU/MI), who conduct research projects, improve infrastructure, and intern with the Air Force Research Laboratory in support of the Air Force mission.

The Air Force also leverages the National Defense Education Program (NDEP) Science Mathematics and Research for Transformation (SMART) Program that supports U.S. undergraduate and graduate students pursuing degrees in 19 STEM disciplines. The Air Force provides advisors for the SMART scholars, summer internships, and post-graduation employment opportunities. The Air Force has sponsored 523 SMART scholars during the past eight years, and of the 315 scholars that have completed the program, 88% are still working for the Air Force, 9% are getting advanced degrees, and 3% have left due to various reasons including furlough and government funding uncertainty. The Air Force identified 110 Key Technology Areas essential for current and future support to the war fighter, which we used for selecting academic specialties for SMART scholars. SMART Scholars are an essential recruitment source of employees to enable key technology advances and future STEM leaders.

Sequestration and fiscal uncertainty in Fiscal Year 2013 caused the Air Force to significantly curtail travel expenses and severely limit conference attendance. It is essential for our scientists and engineers to be fully engaged within the national and international community so this curtailment disproportionately impacted the S&T community. We have worked with Air Force leadership to solve these issues and establish policies allowing greater flexibility for this mission imperative in 2014 and beyond. We can recover from the one year (2013) of non-participation in the greater S&T national and international community. However, severe travel restrictions over the long term could undermine the Air Force's ability retain top talent.

The Air Force has effectively used the authority provided by Section 219 of the Duncan Hunter National Defense Authorization Act not only to increase the rate of innovation and accelerate the development and fielding of needed military capabilities but also to grow and develop the workforce and provide premier Laboratory infrastructure. For example, the Information Directorate of the Air Force Research Laboratory located in Rome, New York used funding made available by Section 219 to develop curriculum at Clarkson University. The curriculum is aligned to the Information Directorate's command, control, communications, cyber and intelligence (C4I) technology mission and provides training and development programs to Laboratory personnel. To fully utilize the new Section 219 authorities from the Fiscal Year 2014 National Defense Authorization Act, the Laboratory is now developing a targeted infrastructure plan to provide its scientist and engineer workforce premier laboratory facilities in its locations nationwide. Recent success in the infrastructure area includes the opening of two state-of-the-art fuze laboratories at Eglin AFB, Florida, which are enabling enhanced research and development into hardened penetration and point burst fuzing.

CONCLUSION

The threats our Nation faces today and those forecast in the future leave the U.S. military with one imperative. We must maintain decisive technological advantage. We must take lessons from the last decade of conflict and creatively visualize the future strategic landscape. We must capitalize on the opportunities found within this space.

The focused and balanced investments of the Air Force Fiscal Year 2015 S&T Program are hedges against the unpredictable future and provide pathways to this flexible, precise and lethal force at a relatively low cost in in relation to the return on investment. We recognize that fiscal challenges will not disappear tomorrow, and that is why we have continued to improve our processes to make better investment decisions and efficiently deliver capability to our warfighters.

Chairman Thornberry, Ranking Member Langevin, Members of the Subcommittee and Staff, thank you again for the opportunity to testify today and thank you for your continuing support of the Air Force S&T Program.

www.ingramcontent.com/pod-product-compliance
Lightning Source LLC
Chambersburg PA
CBHW081814170526
45167CB00008B/3439